Guess Who
Spins

Sharon Gordon

BENCHMARK **B**OOKS

MARSHALL CAVENDISH
NEW YORK

I can live inside or outside.

But you may not see me.

My body has two parts.

My eyes, legs, and mouth are on the top.

My bottom part is large and round.

I have eight legs.

If I lose one, I grow
a new one.

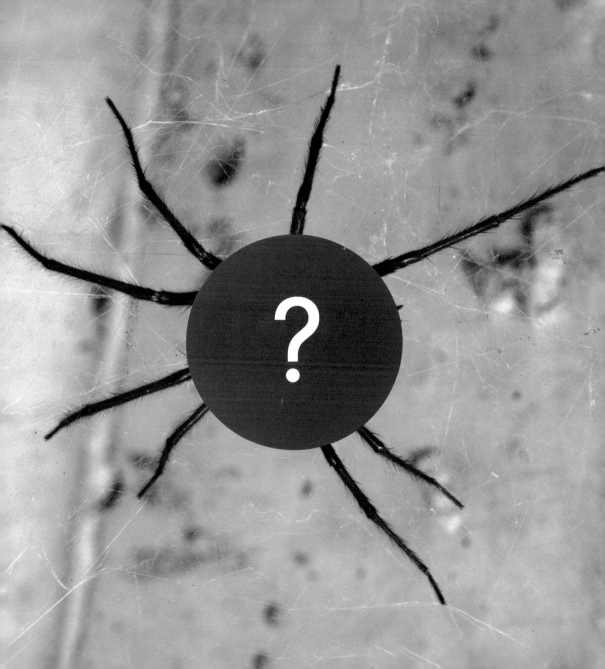

Each leg has small claws
at the end.

These claws help me
climb.

I can walk upside down!

I have two *feelers*.

I use them to eat my food.

Little hairs are on my body.

They feel the air move
when something is near.

I have two strong jaws.

Each one has a sharp
fang.

I can spin a web.

It is made of *silk*.

It is strong and sticky.

I catch my food in the web.

Sometimes I eat it right away.

Sometimes I save it for later.

I lay small eggs.

I wrap them in silk.

My babies are born
in a few weeks.

They live for about
one year.

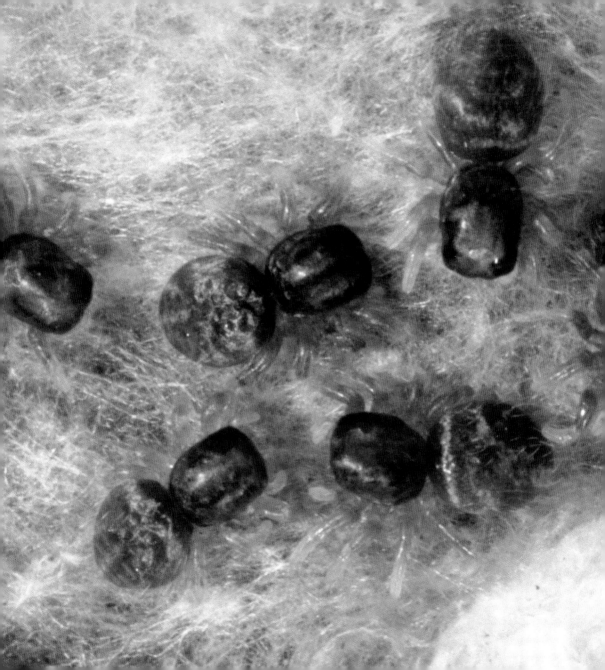

I am getting closer.

Do you see me now?

Who am I?

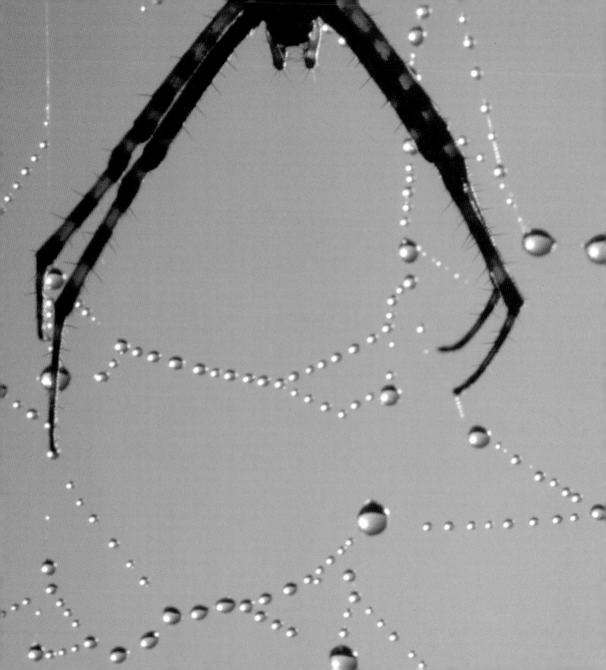

I am a spider!

Who am I?

claws

eggs

fangs

feelers

28

legs   **web**

Challenge Words

fangs

The parts of a spider that grab and hold food.

feelers

The two parts by a spider's mouth that it uses to eat.

silk

The sticky string spiders use to spin webs.

29

Index

Page numbers in **boldface** are illustrations.

About the Author

Sharon Gordon has written many books for young children. She has always worked as an editor. Sharon and her husband Bruce have three children, Douglas, Katie, and Laura, and one spoiled pooch, Samantha. They live in Midland Park, New Jersey.

With thanks to Nanci Vargus, Ed.D. and
Beth Walker Gambro, reading consultants

Benchmark Books
Marshall Cavendish
99 White Plains Road
Tarrytown, New York 10591-9001
www.marshallcavendish.com

Library of Congress Cataloging-in-Publication Data

Gordon, Sharon.
Guess who spins / by Sharon Gordon.
p. cm. — (Bookworms: Guess who)
Includes index.
ISBN 0-7614-1768-0
1. Spiders—Juvenile literature. I. Title II. Series: Gordon, Sharon. Bookworms: Guess who.

QL458.4.G67 2004
595.4'4—dc22
2004003409

Photo Research by Anne Burns Images

Cover Photo by *Corbis*/Robert Pickett

The photographs in this book are used with permission and through the courtesy of:
Corbis: pp. 1, 13 Joe McDonald; pp. 9, 28 (top l.) Clouds Hill Imaging Ltd; pp. 17, 29 (right)
Julie Habel; p. 25 Dewitt Jones. *Animals Animals*: p. 3 James Robinson. Dwight R. Kuhn: pp. 5, 11,
15, 23, 28 (bottom l.& r.). *Peter Arnold*: pp. 7, 21, 27, 28 (top r.), 29 (left) Hans Pfletschinger.
Visuals Unlimited: p. 19 Steve Maslowski.

Series design by Becky Terhune

Printed in China
1 3 5 6 4 2